The research and publication of this book was supported in part by a Graduate Student Research Grant from Organizational Dynamics, School of Arts and Sciences, University of Pennsylvania

O N E

"We didn't start the fire; it was always burning since the world's been turning."

Billy Joel

The term "sustainable," and what consumers assume are its aliases—natural, local, and organic, is being used in an unsustainable way. For every different or watered-down version of this definition, consumers become confused, frustrated and at worst, jaded. The first step in reversing this cynicism (before it turns to apathy) is determining an agreed-upon definition of sustainable and then providing people with viable applications. The Natural Step Framework, a proven scientific framework that is internationally recognized, is a logical choice. However, most Americans are not familiar with this framework.

An easy way to interest and educate our citizens on the manifestation of sustainability, as well as the strategy behind it (important to combat green washing), is by teaching them the process by which food reaches their tables. Everyone eats everyday. American culture revolves around food—and lots of it. From sporting events to celebrations to bookstore cafés, food is everywhere. While most Americans can relate to food, they are unaware of the extent that their food choices impact themselves, their communities, their country, and their planet. Cloth shopping bag sales are up, hybrids are chic, and yet, paying ten cents more for a local apple versus buying a box of Apple Jacks seems like an indulgence—despite our current industrialized food system contributing as much as 37 percent of the green house gases in the United States and using 19 percent of the nation's fossil fuels.[1]

While Americans love a bargain, evidence shows many are willing to invest in their short and long-term health. Organic food sales have more than quintupled, increasing from $3.6 billion in 1997 to $21.1 billion in 2008.[21] Food sold by farmers directly to household consumers rose 49 percent to $1.2 billion in 2007 from $812 million in 2002.[22] These are signs trending in the opposite direction of the current food system, which mounting evidence indicates is unsustainable to our country's health and environment. But much of the infrastructure, even with organic foods, is built around a system that already isn't working and relies heavily on oil. So, what is the answer? Is there one or even ten solutions?

The purpose of this book is to analyze some of these issues and raise necessary questions in the hope that U.S. consumers can better understand the positive aspects of sustainable food supply—for both themselves and the environment—by using the following strategy:

- Deepening the awareness around the impact of their food choices
- Providing general sustainable food choice guidelines and resources bound by the Natural Step Framework
- Considering the true sustainability of a new food system and choices

Chapter 1:

"We didn't start the fire; it was always burning since the world's been turning."

~Billy Joel

Chapter 2:

"No problem can be solved from the same level of consciousness that created it."

~Albert Einstein

Chapter 3:

"You can't always get what you want. But if you try, sometimes you might find you get what you need."

~Rolling Stones

Chapter 4:

"In the book of life, the answers aren't in the back."

~Charlie Brown

End Notes:

Table of Contents

Global warming. National security. A crumbling health-care system. An energy crisis—again. Don't forget about swine flu! And that's just the first page of your online newspaper. However, being an optimist, you remind yourself that in times like these, there have always been times like these. So you buy a hybrid or bike to work, invest in a cloth shopping bag to put your organic food in and/or recycle religiously.

You feel a bit smug when you see an SUV on the highway or (gasp!) your neighbor still using plastic bags. But as you take out your recycling container, the one with all the cereal boxes and plastic food containers, do you realize you hold something that could not only help the environment more than any other single action you've previously taken but also solve the problems of the world that are making your temperature, along with the Earth's, rise?

Yes, Virginia, there is a Santa Claus, and he's slimming down by eating locally grown organic food. In addition, if you join him, you too can save yourself, your community, your country, and your world. It's true!

It's time to turn your devilish pitchfork into a peace fork. Every time you eat, whether that is three or six times a day, you are voting for the kind of world that you want to live in. Yes, that's quite a mouthful. Chewing it isn't always what you'd expect!

Let's take a look at how that fork is changing more than your waistline. On the next page is a visual model of how our current industrial food system is "working." Please note that the nuances in each part of the cycle are infinite. For example, the health-care collapse is partly caused by overuse of the system from both preventable and current diseases caused by over-testing which is caused by litigation which is caused by misdiagnosis which is caused by variation in doctors' education which is caused by pharmaceutical companies which leads to limited biased research and so on. You get the picture.

However, this is an attempt at providing a visualization of how this process works. In the next chapter, I'll explore these problems as well as provide proposals for more sustainable decisions.

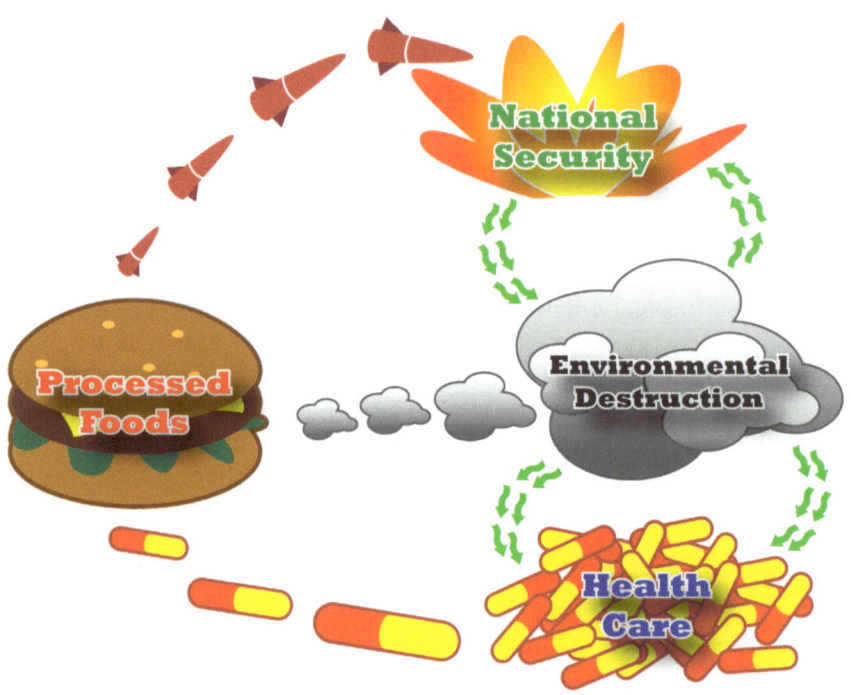

The focal point for this destruction, both ways, is processed foods. Let's start by going up. Currently, it takes 10 calories of fossil fuel energy to produce a single calorie of modern supermarket food, up from 2.3 calories in 1940. Translation for us non-science people: pass the oil to drizzle on your cereal—and the rest of your grocery cart. While you fuel your carbohydrate addiction, you are fueling America's oil addiction.

The dependence on oil for our national food system enriches oil-rich countries like Saudi Arabia and Iran. This resource contributes to their wealth, power, clout, and ability to aid terrorist groups, threatening U.S. national security. In addition, the majority of industrial meat and other processed food come from a handful of corporate processing plants. This concentration makes attacking our food supply simple.

In December 2004, outgoing Secretary of Health and Human Services Tommy Thompson said, "For the life of me, I cannot understand why the terrorists have not attacked our food supply because it is so easy to do," adding that he "worried every single night" about terrorist threats to the nation's food supply.[2]

Terrorist attacks, the ensuing military casualties, sense of insecurity, and other consequences create substantial long-term stress on a country and its citizens, resulting in increased use of the medical system. Stress alone is enough to burden the system. According to a special report by the New England Journal of Medicine in November 2001, because of the September 11 terrorist attacks, Americans across the country, including children, suffered substantial symptoms of stress: 44 percent of adults studied reported one or more substantial symptoms of stress and 90 percent experienced one or more symptoms to at least some degree.[3]

Flying with raised color alerts, periodic false alarms, and terrorist attacks all increase stress. A litany of health problems from heart disease to obesity to depression can be linked to stress. This increases the need and usage of the U.S. health-care system, which relies heavily on pharmaceutical drugs. These, along with other medical materials like rubber gloves and intravenous tubing, are mostly made from petroleum derivatives. In addition, according to a 2008 *Time* magazine article, square foot by square foot, hospitals use twice as much energy as office buildings. Health care is the second most energy-intensive industry in the U.S. (after, you guessed it, food service and sales) with energy costs of $6.5 billion a year—a number that continues to rise right along with the Baby Boomers' age.[4] I bet those stats don't include the energy used to clean up those drugs in the water supply that the Associated Press found in at least 41 million Americans' water.[5]

All these facts make something abundantly clear: more status quo or western health care equals more global warming. Side effects of global warming include rising sea levels and diminishing crop yields due to declining rain fall levels. This will cause more than one billion people, mostly in developing

> "FOR THE LIFE OF ME, I CANNOT UNDERSTAND WHY THE TERRORISTS HAVE NOT ATTACKED OUR FOOD SUPPLY BECAUSE IT IS SO EASY TO DO"
>
> TOMMY THOMPSON
> - FORMER U.S. SECRETARY OF HEALTH AND HUMAN SERVICES

nations, to scramble to find a new home by 2050.[6] With shelter a higher survival priority than food quality, the immediate cheap lure of processed foods will continue. Seems Darwin and his survival of the fittest theory is no joke!

Deep breath. What goes up must come down. It's time to set the previous diagram into reverse by connecting how processed foods lead to environmental destruction. Processed foods rely heavily on fossil fuels and pesticide use for their production. Four pounds of pesticides are used for every U.S. man, woman, and child every year.[7] Processed food and its consumption are responsible for about 37 percent of all greenhouse gases and 19 percent of fossil fuel consumption.[1] That sort of energy use seems considerably inefficient and bizarre when the unlimited resources of the sun can provide most of the energy needed to grow real food (i.e. vegetables, fruits, nuts, seeds, beans, grains) through the process of photosynthesis.

Meanwhile, the steady effects of a warming climate, epidemiologists say, will lead to an increase in infectious and chronic conditions, such as allergies and respiratory disease.[6] Air pollution from fossil fuel use at its least lethal level contributes to these diseases as well as skin and eye irritation, and at maximum danger, a variety of cancers. Pesticides are associated with birth defects, nerve damage, cancer, and "other effects that might occur over a long period of time." And, this is directly from the conservative Environmental Protection Agency which has numerous members who have previously worked for chemical companies!

Again, we have an overuse of the health-care system, which as we are currently experiencing here in the U.S., means not everyone has financially viable options for care and even those who do, experience considerable financial loss. For example, a 2007 study found that 62 percent of all bankruptcies filed in 2007 were linked to medical expenses. Of those who filed for bankruptcy, nearly 80 percent had health insurance.[8] This burden creates a system where not everyone can have basic needs met. This inevitability targets the already economically challenged and further divides the line between the poor and struggling middle-class with the rest of society. An increase in poverty leads to more crime which is a domestic issue for national security. An increased indigent population makes it more difficult to command a fair price for food and increases dependence on cheap (albeit government subsidized) processed foods.

There is no need to go into how poor quality food is affecting the current health care system. I'll let the Centers for Disease Control and Prevention sum it up: Of the $2 trillion we're spending on healthcare, $1.5 trillion is for the treatment of *preventable* chronic disease. Four of the top 10 killers in America today are chronic diseases linked to diet: heart disease, stroke, Type 2 diabetes, and cancer.[9] I know a country that could use an extra $1.5 trillion right about now.

It's Armageddon. That's right, the weapons of mass destruction were never in Iraq. They are at your local fast food joint, in your freezer, and (gasp!) even at Whole Foods. But the great news is that if you can snap out of your food coma, you'll have the energy and stamina to do your part. And while you are saving the world, you'll also save yourself.

Here's what you can look forward to when you eat in a sustainable way:

- Svelte figure without dieting

- Energy levels that make you insanely productive and accomplished

- No allergies, colds or doctors visits to slow you down

- Bagless eyes because you sleep so well

- Clear skin

- A more relaxed "you" who isn't hyped up on stimulants like high-fructose corn syrup and food chemical additives

- Confidence in knowing the source of your food while feeling connected to your community

- Preventing and reversing disease

- Satisfaction from contributing to a more just, peaceful, and cleaner world where people are paid fairly while not being exposed to chemicals you would never want your family to touch

Okay, you're in. It's a no-brainer, really, you think, and confidently place your order for real and sustainable food. Yet again, however, this only leads to another essential question: *What is real, sustainable food?*

T W O

"No problem can
be solved from
the same level
of consciousness
that created it."

Albert Einstein

To explain guidelines for sustainable food, it's important to define both food and sustainability. Both words get used liberally these days and even with established definitions, the answers aren't simple.

Food: *an edible nutrition source that isn't advertised or pre-packaged. Usually contains one ingredient, sometimes two but rarely more than five (unless in a recipe).*

When you think of your grocery cart, how much of it is stocked with food (fruits, vegetables, nuts, seeds, beans, whole grains, fish, meat, eggs)? How much of it is in boxes? How much of it is chemicals? How much of it can you pronounce?

Sustainability: *those developments that meet present needs without compromising the ability of future generations to meet their needs* (as defined by the United Nations in 1987).

An easy way to think of this is by following the golden rule but on a generational level: do unto your kids and grandkids what you'd like them to do unto you (you know, create a clean environment where you are mentally and physically agile without being on 20 medications and have a nice, safe, non-toxic home with lots of friends and family).

Because we haven't been thinking this way, we've gotten ourselves into a funnel. There are increasing demands with decreasing resources.

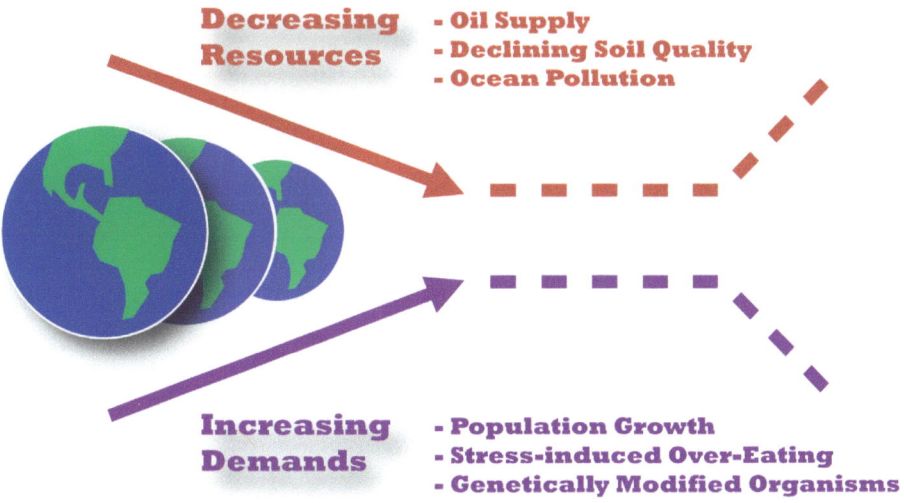

Decreasing Resources
- Oil Supply
- Declining Soil Quality
- Ocean Pollution

Increasing Demands
- Population Growth
- Stress-induced Over-Eating
- Genetically Modified Organisms

Decreasing Resources:

- Oil supply for food production

- Declining soil quality due to industrial farming practices and air quality

- Changing temperatures (global warming) leading to erratic rainfall causing crop failures

- Ocean pollution and over-fishing

- Pesticides, herbicides, and genetically modified foods lowering crop yield

Increasing Demands:

- Larger food supply because of population growth

- Over-eating due to stress, emotions, and malnutrition fueled by a processed food diet

- Use of Genetically Modified Organisms

The further we get into the funnel, the less room we have for error. Considering how well we are doing now, I'm betting on the House (the Earth) and Her solution to do away with most of us if we don't shape up. It's important to note that the funnel is not our destiny. We don't have to be swallowed alive. It is the product of our policies, practices, and patterns of behavior. We have the power to not only change the shape of the funnel, but stay out of it all together. There are still opportunities to get out of the funnel and back into Her good graces. Is it by buying organic? Or is it buying "local"? How is "local" defined? How about buying from small-scale artesian producers from Foodzie. com? What do all these terms mean anyway? Just like Enron, Florida voting machines, and the

U.S. housing market, not everything is as it seems. Thinking "sustainably" requires a completely different way of viewing every food purchase you make. Many of the recommendations in this chapter are only the beginnings of converting to a more sustainable food culture. They are by no means a final answer, as any change to a system creates entirely new conditions and must be monitored and evaluated continuously.

The guidelines offered in this book are based upon The Natural Step (TNS), a non-profit organization founded with the vision of creating a sustainable society. It was started by Dr. Karl-Henrik Robert in the late 1980's. He was an oncologist working with a large number of children who were suffering from cancer. He was struck by the need for gaining an understanding of cells, and therefore of the continuation of life. He developed the framework outlining the essential conditions for life. All types of groups from government to businesses concurred on these conditions. Sweden, his native country, was so intent on implementing the tenants of TNS, it printed and distributed its framework to every citizen of its country—more than 20 years ago.

For two decades, TNS has been at the forefront of international research and dialogue about sustainable development. As a result, they have developed a proven framework that helps us scientifically judge whether our decisions and actions are truly sustainable versus merely shifting the burden to the next generation.

TNS provides four conditions by which to judge our actions in order to determine if we are reducing the impact on the Earth, not exacerbating it. By understanding this framework, you can cut through all the "green-washing" and companies claiming to be green, sustainable or whatever trendy words corporate branding is using to market to you and other environmentally conscious consumers.

According to TNS, to improve the Earth as a system, including the people on it, become a sustainable society, and prevent our own self-destruction, we must follow their precepts.

THE NATURAL STEP FRAMEWORK FOR SUSTAINABILITY

1. Eliminate our contribution to the progressive buildup of substances extracted from the Earth's crust (such as fossil fuels from shipping food from around the world, less petroleum-based pharmaceutical use).

2. Eliminate our contribution to the progressive buildup of chemicals and compounds produced by society (such as fertilizers and pesticides on food, less BPA in cans, jars and lids).

3. Eliminate our contribution to the progressive physical degradation and destruction of nature and natural processes (such as soil degradation from industrial farming and over fishing the ocean).

4. Eliminate our contribution to conditions that undermine people's capacity to meet their own basic human needs (for example, unsafe working conditions and not enough pay to live on). Humanity's basic human needs were identified by Chilean economist, Manfred Max-Neef: subsistence, protection, affection, idleness, identity, freedom, creativity, participation, and understanding[24].

When making new decisions, a return on investment (ROI) is expected. Within TNS, ROI can mean financial, political, social or other types of return. And it's not that we pretend our new decisions will eliminate all the challenges from the previous strategy. The idea is to improve upon them as much as possible.

To illustrate this example, let's look at the cancer industry. Breast cancer rates continue to rise, indicating a need for a system improvement. Companies like Mars, Inc. who makes M&M's flaunt pink ribbons and provide sponsorship money, yet scientific evidence shows that sugar and insulin surges fuel breast cancer.[10] Additionally, pink ribbon clothing contributes to the use of herbicides and pesticides—cotton is one of the most heavily sprayed crops in the world. Don't get me started on all of the BPA-containing pink plastic products that are sold to support breast cancer awareness despite the plastic-breast cancer connection![11]

A more sustainable solution, in reaching the goal of eliminating breast cancer (always think of the end goal and work backwards), is to use the TNS framework which utilizes the four system conditions. Ideas like partnering with environmental groups and local, organic farmers who are working on preventing cancer by producing nutritious food and reducing the pesticides/chemicals in our environment (more than 70 percent of breast cancers are linked to environmental factors) is a more viable option.[11] Offering awareness materials from organic cotton would positively correlate with the system conditions of TNS.

"COMPANIES LIKE MARS, INC. WHO MAKE M&M'S FLAUNT PINK RIBBONS AND PROVIDE SPONSORSHIP MONEY, YET SCIENTIFIC EVIDENCE SHOWS THAT SUGAR AND INSULIN SURGES FUEL BREAST CANCER."

At the very least, this sustainable option utilizes all of TNS conditions. There is a reduction of toxic materials in the environment. Elimination of pesticides and herbicides through organic farming halts the physical degradation of natural processes and contributes to the well-being of the cotton-field workers. The net result produces a great ROI. Here, you have a financial and social ROI. At the very least, fewer women will have to deal with the incredible burden of cancer; good health is invaluable to them, their families, and the medical system.

The guidelines in the next chapter are by no means absolute and without debate. They are an attempt to start moving us in the direction of a world where every human being has access to healthy food that is produced in a clean environment that enables healthy bodies to live prosperous lives. They are the low-hanging fruit. The stuff higher in the trees will be addressed, but left open to your creative imagination to solve—helping you and generations to come.

THREE

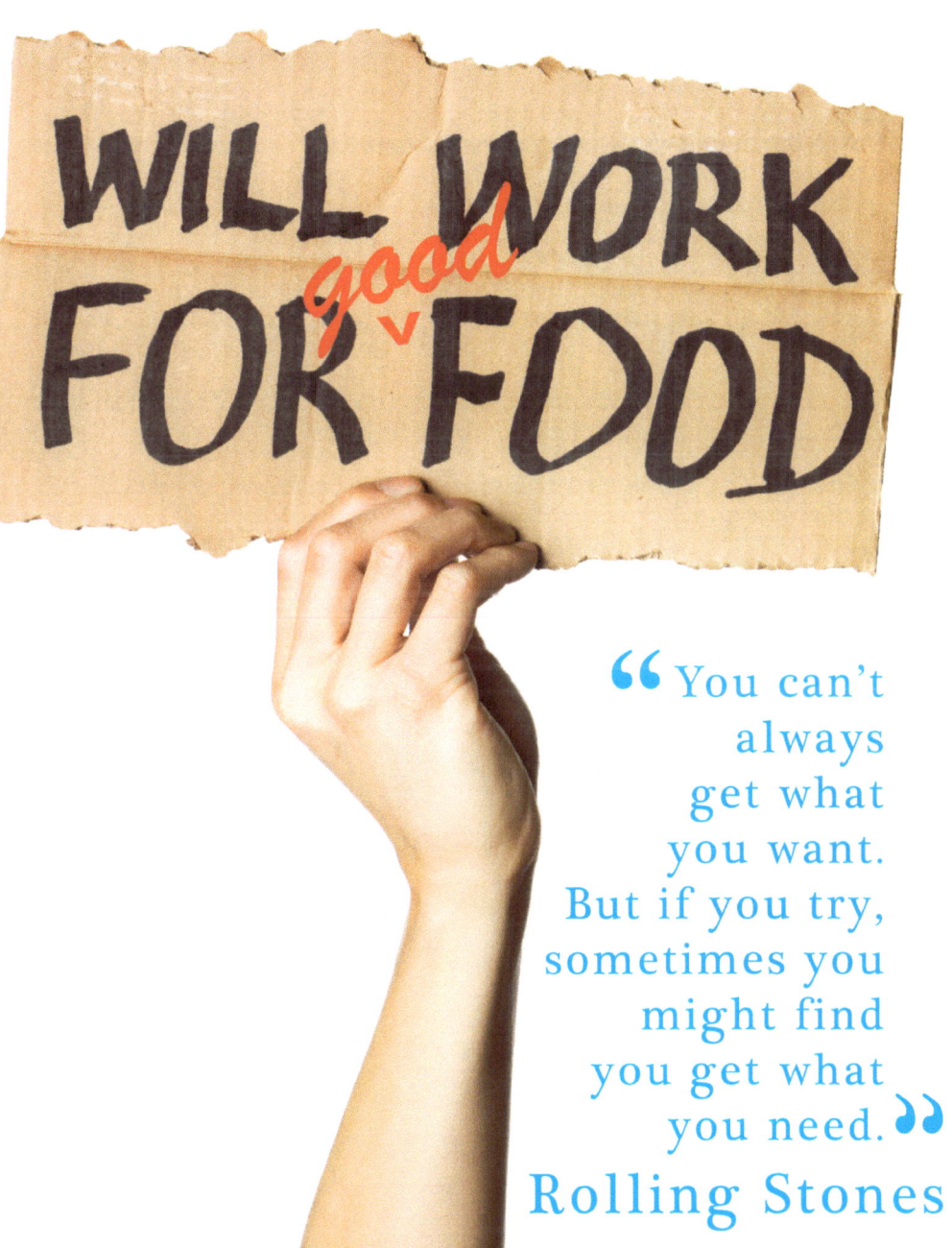

WILL WORK FOR good FOOD

"You can't always get what you want. But if you try, sometimes you might find you get what you need."

Rolling Stones

Wouldn't it be lovely if all the answers to our global food crises could be wrapped up nicely in a biodegradable take-out container?

Some of the answers are ready to go and they are take-out ready in this chapter. Keep in mind, though, that all systems—even local and organic farming—have their challenges (explored in the next chapter). The goal here is to address and relate the four system conditions as much as possible. Think progress, not perfection. The questions that will be posed in chapter four will hopefully generate a conversation around the unanswered questions and provide a spark for some creative solutions.

Five Strategies to Sustain Your Smug Smile

1. Eat Local and Organic Food

From the Nicoyans of Costa Rica to the Okinawans in Japan, the healthiest and longest living people on earth eat very different diets but have two things in common: they eat real food that comes from their land and not a box and they don't spend nearly as much as the U.S. does on health care.

There are several ways eating real food grown locally and organically also lessens the negative impact of the food industry on the environment:

TNS-1: *Eliminate our contribution to the progressive buildup of substances extracted from the Earth's crust.* There is less fossil fuel usage from production, packaging, and transportation. Real food's major energy input is the sun, an unlimited energy resource, versus the finite sources of oil, natural gas, etc.

TNS-2: *Eliminate our contribution to the progressive buildup of chemicals and compounds produced by society.*

Organic means avoiding dangerous pesticides and fertilizers. Local, real food also requires less packaging such as canned and glass jars, eliminating toxins such as BPA (Biphenol A), a hormone disrupter known to contribute to infertility and miscarriages, and breast cancer.[12]

TNS-3: *Eliminate our contribution to the progressive physical degradation and destruction of nature and natural processes.*
Local and organic farming dramatically improves biodiversity and crop yield while minimizing carbon outputs. This creates a cleaner, richer soil. Many industrial farms are mono-culture farms, which stress the land more than multi-crop farms and subsequently require more chemical use. Certain crop loss due to pests was double prior to chemical use.[7]

TO FIND LOCAL AND ORGANIC PRODUCE NEAR YOU, GO TO LOCALHARVEST.ORG

TNS-4: *Eliminate our contribution to conditions that undermine people's capacity to meet basic human needs.* Buying locally affords farmers a fair price for their labor. This adds to the basic human needs of subsistence (more nutrition), protection (stronger local economy, safer food), affection (connecting with community, freedom (being healthy is the biggest freedom of all), and participation (voting for a community you believe in).

Ringing Up the Sale:

- Grow your own—start with low-maintenance herbs

- Join a Community Supported Agriculture (CSA): visit local harvest.org for one near you.

- Find a farmers market near you—localharvest.org

- Shop at grocery stores that source from local producers

- Support restaurants that source from local producers

For the Crying Baby in the Buggy:

Problem: "I don't have access to local and organic."

Solution: Go as local as possible. As Erin Barnett, Director of Local Harvest, said, "Local is just a shorthand for quality. It's more than just a passing glance on stewardship. Family-scale farmers tend to have a long history with their land and take good care of it. It's like having a family pet." Buy your spinach from California not China (note: many frozen vegetables, even from Whole Foods, come from China. Look on the back of the bag for the country of origin).

Problem: "There aren't farmers markets or CSA's near me."

Solution: Again, go as local as possible or check out localharvest.org or foodzie.com for local producers that can ship to your house and are as close to you as possible.

2. Eat Less Meat and When You Do, Purchase Free-Range/Grass Fed.

Indigenous cultures, who didn't understand "Western" disease until the Western diet came to their world, were some of the healthiest people on Earth. Healthy is defined as free of disease. Many of them ate a variety of wild game as a diet staple. Contrary to some judgmental vegans, meat isn't necessarily unhealthy and many people require it for their nutritional needs. It's the quantity and how it's produced that causes sustainability problems.

TNS-1: On average, about 40 calories of fossil fuel energy go into every calorie of feed-lot beef in the United States.[13] Grass-fed meat significantly cuts down on fossil fuel usage, as a bushel of grain takes approximately a half gallon of fuel versus the sun providing the majority of energy needed to grow grass[9].

TNS-2: Because factory farm meat is produced with animals so close to each other, massive amounts of antibiotics are needed. This adds to antibiotic resistance which creates stronger strains of bacteria and pollution in our water streams where antibiotic traces and fecal matter (which on pasture raised farms is used as fertilizer) are destroying ocean life, a major source of food for the world.[14]

TNS-3: The majority of conventional meat comes from Concentrated Animal Feeding Operations (CAFO's), which has most likely influenced Swine Flu.[24] These obviously affect humans but also interfere with the balance within nature's biodynamic. Also, the watering needs of livestock are tremendous, far above those of vegetables or grains. An estimated 1,800 to 2,500 gallons of water go into a single pound of beef.[15]

TNS-4: Many of these factory farms, as documented in *Fast Food Nation* and *Food, Inc.* use unfair labor practices that not only pay less than fair wages, but are physically dangerous and emotionally draining. If social causes aren't your thing, grass-fed and humanely grown meat tastes so damn good. Move over, arugula; this is the "it" item for the elitist gourmet.

Ringing Up the Sale:

- 🍅 Join the Meatless Monday brigade (meatlessmonday.com). Once a week, commit to a dinner without meat. Portabella mushrooms are a great substitute.

- 🍅 Add in more plant-based foods (fruits, veggies, nuts, seeds, beans). They will keep you fuller longer with less meat.

- 🍅 Purchase grass-fed meat from local farmers markets, grocery stores or small farms from localharvest.org.

- 🍅 Support restaurants that buy grass-fed meat.

For the Crying Baby in the Buggy:

Problem: "I can't afford grass-fed meat."

Solution: Use less meat than called for in recipes or pool your resources together to buy a cow or other farm animals. This is very popular in Pennsylvania and New Jersey and is commonplace around the country. You can find a local rancher to purchase from by asking around at farmers markets, contacting localharvest.org or talking to your local butcher.

Problem: "I can't find grass-fed meat."

Solution: Aim for organic meats or find a source on localharvest.org. Organic meat means the animal had access to pasture, wasn't given antibiotics and/or hormones, and has organic feed. Many times its still grain fed but it's better than conventional meat.

3. Avoid Genetically Modified Organisms (GMO) Food

GMOs are engineered seeds whose genetic blueprint is permanently altered. By modifying a seed's hereditary makeup, scientists hope that a plant grown from the seed, and its descendants forever, will have certain character- istics.

While engineering seeds has been going on for centuries, it has been with characteristics of plants already present in the species. These days, companies are trying to cross breed characteristics. For example, a strawberry that had a gene from an arctic fish was inserted into it to make the strawberry more frost-resistant.[16] The short-term and long-term effects of this are undetermined. And yet while 85 percent of Americans oppose GMO foods, these companies have fought to have labeling laws work in their favor, and now most Americans unknowingly are consuming GMOs on a regular basis. Europeans are not consuming as much, as GMOs are banned throughout much of Europe.

Another sustainability hurdle with GMO seeds is they are engineered to withstand the herbicides many of these GMO companies produce (the top five biotech companies that account for nearly 100 percent of the genetically engineered seeds also account for about 60 percent of the global pesticide market.)[16] This means considerably more herbicide use as GMO crops can be sprayed before and during planting (conventional plants cannot be sprayed during the growing season because the herbicides will kill the crops).

This enables more use of certain herbicides. For example, the heavily used Roundup, an herbicide manufactured by Monsanto, one of the most powerful GMO companies in the world, has been linked to loss of wildlife and pesticide-related illness that includes everything from eye and skin irritation to vomiting and non-Hodgkin's lymphoma.[16]

The GMO food debate can fill an entire library. For the sustainability debate, GMOs need to be viewed from a scientific basis rather than a profit motive as Monsanto claims GMOs are sustainable too. Apparently they haven't considered the Natural Step Framework in evaluating their own claims.

If you look at the four system conditions used by the Natural Step, it is clear that America needs to "round up" Monsato for extortion (and bribery for which they have also been convicted).

So let's enlighten Monsanto by relating GMOs to the TNS conditions that are used to sustain life, not profits. The resulting violations of the natural steps are as follows:

TNS-2: Increased use of herbicides

TNS-3: Substantial loss of biodiversity, the foundation of species survival, since GMO crops have the same DNA

TNS-3: Loss of wildlife, including earthworms necessary to soil quality and ecosystem balance, due to pesticide exposure and decreased soil quality.

TNS-4: Loss of farms and livelihood when GMO crops, which can cost as much as twice as conventional crops, fail. For example, in 2003, 17,107 farmers in India committed suicide.[17] While these suicides cannot all be directly attributed to GMO seeds, an increased amount of debt and low yields created a system condition that contributed to the suicide rates.

19

FOR A FULL-LIST OF
GMO FOOD, VISIT
WWW.GEACTION.ORG/
TRUEFOOD/
SHOPPERSGUIDE

TNS-4: What many GMO companies view as weeds to be eradicated, many poor people use for nutrition and medicine, according to Dr. Vandan Shiva, director of India's Research Foundation for Science, Technology, and Resource Policy.

TNS-4: Significantly more health problems from increased herbicide exposure to workers and community. In addition, many GMO companies claim these types of seeds are our ticket out of world hunger, which could ease condition four of the Natural Step Framework. Yet in his special report, *Failure to Yield*, Doug Gurian-Sherman, a senior scientist in the Food & Environment Program at the Union of Concerned Scientists, reports that despite 20 years of research and 13 years of commercialization, genetic engineering has failed to significantly increase U.S. crop yields. "If we are going to make headway in combating hunger due to overpopulation and climate change, we will need to increase crop yields," said Gurian-Sherman. "Traditional breeding outperforms genetic engineering hands down."[18]

So while Monsanto, the leading maker of GMOs has positioned themselves as "sustainable" agriculture on their website "produce more, conserve more, improve farmers lives," this message is clearly "green-washed."

Full disclosure note: Because of damning research not included here for purposes of simplicity, I do love to pick on Monsanto, because it is so easy. But many companies slap the "sustainable" or "green" label on their products—not because it's true but because it will sell. It's why the Natural Step Framework is vital to understand. There will never be a PR person getting one over on you!

Ringing Up the Sale:

- Avoid brands made from GMOs: ConAgra, Heinz, General Mills, Keebler, Kellogg, Quaker, Nabisco, Nestle, Mars, (for complete company and product specific information, log onto: http://www.geaction.org/truefood/shoppersguide).

- Support companies that don't use GMOs like Amy's, Arrowhead Mills, Bob's Red Mill, Pamela's, Endangered Species Chocolate Chips, Health Valley, and Whole Foods 365 brand.

- Support True Food Now, the Organic Consumers Association, the Environmental Working Group, and others committed to keeping sustainable agriculture alive

- Contact your representative about making labeling GMOs a priority

🍅 Watch The Future of Food and/or read *The Food Revolution* by John Robbins.

For the Crying Baby in the Buggy:

Problem: "It seems GMOs are everywhere."

Solution: Remember to buy "food" versus products, or choose organic as the organic label currently doesn't allow GMO seeds.

Solution: Look for bar codes that begin with the number "8" on fruit and produce. This means they are GMO. If you are eating canola oil, corn, or soy, make sure to buy organic. These three crops are the most highly GMO'd crops in America.

4. Buy Fair-Trade Food

It's clear that not everything can be purchased locally. So what can be done to still support a sustainable food supply? Fair trade!

Fair-trade food means food that is fair to the environment, the people, and communities who produce it. As documented in the book, *Grub*, farmers have full price disclosure for how much their food is worth instead of having to adapt to a changing commodities market half-way around the world. One fair-trade farmer from Nicaragua summed up the difference in what being part of the Fair Trade movement meant to her, "Fair trade gives us back our dignity."[7]

According to TransFair USA, a non-profit organization that is the only independent, third-party certifier of Fair Trade products in the U.S., Fair Trade empowers farmers and farm workers to rise above poverty and to develop the business skills necessary to compete in the global market- place. According to its website, "by guaranteeing mini- m u m floor prices and social premiums, Fair Trade enables producers to invest in their farms and communities and protect the environment."

Fair-trade improves system conditions 2, 3 and 4:

TNS-2: While not all foods are produced organically, as sustainable methods as possible are employed. This means less pesticides and herbicides.

TNS-3: As a whole, smaller farms keep the environment cleaner and have higher output. As Dr. Peter Rosset, an agro ecologist and rural development specialist, explains in the *Ecologist*, small farmers are more likely to plant different types of crops, combine or rotate crops and livestock, and use manure to replenish soil fertility.

These integrated farming systems ultimately produce far more per unit area than monocultures, which is what most large farms grow.[19]

TNS-4: The social needs of the farmers and their employees are greatly improved.

Ringing Up the Sale:

- Look for the Fair Trade Certified and Equal Exchange logo—two organizations helping to grow the fair trade market

- Find artisan producers at Foodzie.com and Localharvest.org

- Visit local farmers markets

- Purchase fair-trade products at stores like Starbucks and Wal-mart. For a complete listing, visit: http://www.transfairusa.org/content/WhereToBuy/

5. Choose Local, Except with the Dirty Dozen*

This is where we start climbing the tree. Local, non-organic foods utilize TNS principles 1 and 3, and to a degree numbers 2 and 4. While small farms favor fertilizers like manure and compost versus the agrochemicals used on large farms,[19] they are still used and put farmers and the public at risk for the health hazards associated with chemical use. Too, certain foods require more pesticide use than others. Organic food sometimes is shipped from around the globe, using much fuel in shipment while losing nutritional value by the time of purchase. Many times these workers aren't paid fair wages. Since organic crop farm land only accounts for 0.5 percent of all U.S. cropland, buying organic supports more farmland conversion.[20] Who needs more support right now, local farmers or organic farmland?

"WE LET THE LOCAL FOOD MAKE THE CASE FOR ITSELF. IT'S EXPERIENTIAL. OUR SENSES TELL US THIS IS ALIVE, AND IT'S A BETTER WAY."

ERIN BARNETT
-LOCALHARVEST.ORG

According to the Environmental Working Group, a non-profit designed to protect public health and the environment, people can lower their pesticide exposure by almost 80 percent by avoiding the top twelve most contaminated fruits and vegetables and eating instead the least contaminated. Below is a list of the 2009 Dirty Dozen.

DIRTY DOZEN:

1. Peach
2. Apple
3. Sweet Bell Pepper
4. Celery
5. Nectarine
6. Strawberries
7. Cherries
8. Kale
9. Lettuce
10. Grapes - Imported
11. Carrot
12. Pear

Source: The Enviromental Working Group.

This is where questions TNS provides helps us prioritize our decisions and is the basis for the recommendation. The three questions to ask yourself when things can get a bit dicey:

1. Does this action move us in the right direction (toward alignment with the sustainability principles)?

2. Can this action be built upon in the future? (i.e. is this a flexible platform?)

3. Does this action bring an acceptable financial, ecological and/or social return on investment?

It doesn't mean this is a perfect answer, but eating locally and avoiding the dirty dozen move us towards our end goal; it's flexible and can help evolve a better system in the future (keeping both necessary markets in existence) and brings acceptable ROI in different ways.

Your Change

The synergistic effect of these five actions is that when properly nourished, one eats less and is much healthier. Consuming less is a magic bullet for any food system improvement grounded in sustainability, as defined by TNS framework. It also includes the experience of being connected to your food, cooking and eating more with family and friends, and the general sense of well-being that comes from eating real food versus processed food. As Erin Barnett from Local Harvest said, "We let the local food make the case for itself. It's experiential, our senses tell us this is alive, and it's a better way."

But as a new system of food production arises, new questions are formed to keep us humans busy, with purpose, and without a smug look on our face.

FOUR

"In the book of life, the answers aren't in the back."

Charlie Brown

It's important to feel great about yourself—and eating real food, as fresh as possible, will help that. It's also important to realize even if you start treading a lot lighter on this beautiful world of ours, it's still a dance.

The questions below still circulate like all of those nasty pesticides and food-borne illnesses. As easy as it is to look at our current food system and criticize, it is the product of a series of decisions that probably seemed like a good short-term solution at the time.

Of course you have your corruption, greed, and quest for power, but these problems many times are a result of short-term, "I need to get re-elected next year" or "make my quarterly numbers" thinking. In an effort to authentically look long-term, for which TNS accounts, it's important to acknowledge that in thinking "sustainably," questions still remain.

Special note: these questions are not meant to create confusion and prevent action—as the tobacco industry did for decades to delay the obvious conclusion that cigarettes were bad for our health. Instead, these questions are designed to help consumers tread mindfully and inspire creative solutions that can have a bigger impact.

Here are the questions right above the low-hanging fruit. These concerns need to be considered in depth as the breadth of the new burgeoning local, food system evolves, so the environmental and health burden is not shifted to the next generation. Remember, that's the foundation of sustainability.

Growing Questions Planted Within a Sustainable Food System

Eating Local and Organic Foods:

- How efficient is it for farmers to drive in separately, from hours away, to sell their goods to a growing, but small number of buyers? How do we define and calculate efficiency as delivering quality nutrition versus processed food, an inferior nutrition source?

- How many local farms can feed a region? What is the environmental impact on more people farming?

- What sort of government involvement or shift needs to occur so everyone can afford to pay a fair price for food? How does a new agriculture system feed the needs of a growing population?

- How do you take into account the problem of "too clean hygiene," not being exposed to enough bacteria to allow our natural immunities to function? As one farmer said to me, "I know my raw milk is clean. But how dirty are you?"

- How do we become more connected to our food source so that we are exposed to the critters in rich soil found on properly managed farms?

- Monoculture farms do require less tilling which is an energy intensive process. Does the benefit of local and organic farming offset the extra carbon dioxide emissions?

Eating Less Meat

- What role, if any, should government have in making healthy and environmentally sustainable meat available and affordable?

- How much meat is enough to keep farmers in business and maximize our health?

- What needs to be done to help make the transition from heavily government subsided agriculture to one of sustainability?

GMOs

- How do you keep independent research, the crux of objectivity, alive?

- How do you make companies accountable for long-term effects of their products versus short-term results

- If GMO foods were eliminated, how would it impact the job market?

Buying Fair Trade

- When does the distance the food has to travel negate the benefits of buying Fair Trade?

- Farming is one of the most dangerous occupations in the world. How does one better protect an increasing number of farmers?

- How does government policy shift so U.S. commodities are priced fairly in the U.S. market and abroad as current policies depresses prices in other countries?

Eating Local Except For the Dirty Dozen

- Food quality depends on soil quality which depends upon water, the rain cycle, and weather patterns. Transporting organic food thousands of miles burning fossil fuels ultimately damages soil quality. How do we determine when organic costs more than it gives?

These are just "appetizer" questions that can be best solved using TNS, as new data arises. For some of these questions, the data simply doesn't exist to make an educated and scientific decision. The questions themselves raise concerns that should prompt us to refrain from being smug or complacent. Maybe it's not just about hybrids and cloth bags but as Bill Maher said, "If you care about the planet, it's actually better to eat a salad in a Hummer than a cheeseburger in a Prius."

Intuitively and scientifically, these recommendations make sense. However, at one point, so did a lot of other things humans did. It's important to monitor the new food system, for it too has its challenges. Remember, it's progress, not perfection. After all, if all the world's problems were solved, what would we do with ourselves?

We'll leave that question open until we have to answer it. If simple food choices are made, we may have that chance some day soon.

Simple doesn't mean easy (or convenient), at least at first. However, eating real food is the most intuitive of processes if we are aware of the industrial food system. It involves re-orienting yourself with what you consider food. Use the resources in this book to find local, organic, fair-trade resources for your food. Pick one day a week to skip out on meat. Support brands that don't use GMO food. Start to think not in terms of price, but the long-term cost of your food choices using TNS framework. And most importantly, start cooking and sharing meals with your family and friends.

REMEMBER, THIS REVOLUTION STARTS IN YOUR KITCHEN!

Ali is a certified health counselor and 17-year cancer survivor. She is an honors graduate of the Institute for Integrative Nutrition, which is affiliated with Columbia University's Teachers College. She is also a C.H.E.K certified Health and Lifestyle Coach. Ali completed her undergraduate degree at Penn State University, where she was a Schreyer Honors College scholar. She is currently in a Masters program at the University of Pennsylvania. Her book, "The Roots of Going Green: Your Fork, Your Power" evolved out of her Master's research in sustainability. The book will be used in the curriculm at Penn.

Ali currently resides in Philadelphia, PA where she has built a thriving health coaching practice. She's a regular health contributor on the NBC 10! Show, and has been featured in the Wall Street Journal, Delicious Living Magazine and Philadelphia Women's Journal. In addition, Ali is a professional speaker and workshop presenter in demand throughout the country. She enjoys reading in all of Philly's great parks, yoga, traveling and watching Pittsburgh Steeler football.

For more information, visit
ALISHAPIRO.COM

Complete End Notes

1. Pollen, M. (2008, October 12). Farmer in Chief. The New York Times. Retrieved from: http://www.nytimes.com/2008/10/12/magazine/12policy-t.html

2. Allen, M., Branigin, W., & Mintz, J. (2004, December 3). Tommy Thompson Resigns From HHS, Bush Asks Defense Secretary Rumsfeld to Stay. The Washington Post. Retrieved from: http://www.washingtonpost.com/wp-dyn/articles/A31377-2004Dec3.html

3. Mark A. Schuster, M.D., Ph.D., et al. (2001). A National Survey of Stress Reactions after the September 11, 2001, Terrorist Attacks. New England Journal of Medicine, 345(20), 1. Retrieved from: http://content.nejm.org/cgi/reprint/345/20/1507.pdf

4. Schwartz, J. (2008, November). Putting Health Care on an Energy Diet. Retrieved from: http://www.time.com/time/health/article/0,8599,1857853,00.html

5. Donn, J., Mendoza, M., & Pritchard, J. (March 10, 2008) Pharmaceuticals lurking in U.S. drinking water. Retrieved from: http://www.msnbc.msn.com/id/23503485/

6. Christian Aid. (May 2007) Human tide: the real migration crises. Retrieved from: http://www.christianaid.org.uk/Images/human-tide.pdf

7. Lappe, A. & Terry, B. (2006). Grub: Ideas for An Urban Organic Kitchen. New York: Penguin Book.

8. Himmelstein, D, E., et al, (2009). Medical Bankruptcy in the United States, 2007: Results of a National Study. American Journal of Medicine, 122(8), 741-746. Retrieved from: http://www.amjmed.com/article/S0002-9343%2809%2900404-5/abstract

9. Democracy Now (Producer). (2009, May 14). Omnivore's Dilemma Author Michael Pollan's New Advice on Buying Food: "Don't Buy Any Food You've Ever Seen Advertised". Podcast retrieved from: http://www.democracynow.org

10. Hopkins, V., Lee, J., & Zava, D. (2002). What Your Doctor Might Not Tell You about Breast Cancer: How Hormone Balance Can Save Your Life. New York: Warner Books.

11. Cornell University (2008, June 5). Breast Cancer - The Estrogen Connection: Plastics. Video retrieved from: http://envirocancer.cornell.edu/research/endocrine/videos/plastics.cfm

12. Yale University Public Affairs. (2008). Study Shows Why Synthetic Estrogens Wreak Havoc on Reproductive System. In Health & Medicine | Yale Bulletin. Retrieved from http://opa.yale.edu/news/article.aspx?id=1475

13. Pimentel, D. & M. Pimentel. (2003). Sustainability of meat-based and plant-based diets and the environment. [Electronic version]. American Journal of Clinical Nutrition, 78: p. 66S-3S.

14. Walsch, B. (2009, August 31). America's Food Crisis and How to Fix It, Time, 30-37.

15. Kreith, M. and Davis, C.A. (1991) Water Inputs in California Food Production. Sacramento, CA: Water EducationFoundation.

16. Robbins, J. (2001). The Food Revolution:How Your Diet Can Help Save Your Life and Our World. York Beach: Conari Press.

17. Sengupta, S. (2006, September 19). On India's Farms, a Plague of Suicide. The New York Times. Retrieved from: http://www.nytimes.com/2006/09/19/world/asia/19india.html

18. Doug Gurian-Sherman. (2009). Failure to Yield: Evaluating the Performance of Genetically Engineered Crops. Cambridge, MA: Union of Concerned Scientists.

19. Rosset, P. (1999). Small Is Bountiful. [Electronic version]. The Ecologist, 29;p. 452-457.

20. United States Department of Agriculture. Organic Production [Data file]. Retrieved from http://www.ers.usda.gov/Data/Organic/

21. Greene, C., Dimitri, C., Biing-Hwan, L., McBride, W., Oberholtzer L., Smith, T.(2009, June). Emerging Issues in the U.S. Organic Industry. Retrieved from: www.ers.usda.gov/publications/eib55/eib55fm.pdf.

22. United States Department of Agriculture. (2009). Direct farm sales rising dramatically, new Agriculture Census data show. In Farmers Markets and Local Food Marketing. Retrieved from: http://www.ams.usda.gov/AMSv1.0/ams.fetchTemplateData.do?template=TemplateG&navID=WholesaleandFarmersMakets&leftNav=WholesaleandFarmersMarkets&page=WFMFactsAboutFarmersMarkets&description=Facts%20About%20Farmers%20Markets&acct=frmrdirmkt

23. Kirby, David. (2009. April 26). Swine Flu Outbreak -- Nature Biting Back at Industrial Animal Production?. The Huffington Post. Retrieved from: http://www.huffingtonpost.com/david-kirby/swine-flu-outbreak----nat_b_191408.html

24. Fundamental Human Needs. (n.d.). Retrieved September 30, 2009, from The Free Encyclopedia Wikipedia: http://en.wikipedia.org/wiki/Fundamental_human_needs

www.ingramcontent.com/pod-product-compliance
Lightning Source LLC
Chambersburg PA
CBHW050759290526
45792CB00008B/2244